Solar Photovoltaic Training for Residential, Commercial and Utility Systems

Copyright: Steven Magee 2010

Edition 1

Cover Picture: The sun as viewed through high cirrus

Contents

1. Introduction

Solar photovoltaic technology has now become mainstream and is widely used around the world. All of the items needed to construct a reliable system can now be purchased from many different vendors and standards now exist in the industry to ensure that these different products can be seamlessly integrated together. The future of solar photovoltaics is bright and rapid adoption of the technology is underway.

This book will teach the foundations of solar photovoltaics to those who are new to the field. For those already in the field, it should clear up any misconceptions that may have been formed regarding solar photovoltaics. Solar photovoltaics looks relatively easy to understand on the surface, but once you are involved with it on a large scale you will soon realize that there are many pitfalls to avoid. This book is aimed at making you successful and well informed so that you can be the very best in solar photovoltaics.

Solar photovoltaics is very different from conventional electrical system theory and because of this it has its own electrical codes that are contained in National Electric Code (NEC) Section 690 Solar Photovoltaics in the USA. We will explore the concepts of solar photovoltaics and you will become familiar with the differences between solar photovoltaics and conventional electrical theory.

Work safe, follow the codes and build great systems!

2. The Basics of Solar Photovoltaics

Solar photovoltaics comes in many different types:

- Monocrystalline
- Polycrystalline
- Thin film
- Other technologies

Although the technology is different between each type, they all do the same thing. All direct current (DC) solar modules generate DC electricity when exposed to sunlight. The listing above is in order of how efficiently each type will convert sunlight into electricity.

A typical direct current (DC) module will have the following electrical ratings on its label:

- Temperature adjustments
- DC Open circuit voltage (Voc)
- DC Maximum power point voltage (Vmpp)
- DC Short circuit current (Isc)
- DC Maximum power point current (Impp)
- DC Rated system voltage

All of these values are given for Standard Test Conditions (STC). Lets look at what each one of these mean:

Standard Test Conditions (STC)

Standard Test Conditions (STC) is how the solar module performs at a temperature of 25 °C, an irradiance of 1000 W/m² with an air mass 1.5 (AM1.5) spectrum. This is a standard test for all solar modules that are manufactured for the USA market that was developed by the photovoltaic industry and the government. It represents an average set of conditions that can be expected at the mid point between North and South of the contiguous forty-eight states during spring and fall with the sun perpendicular to the solar module. San Francisco, California and Wichita, Kansas are near this midpoint of 37 degrees latitude. In Asia Seoul, Korea, is near and in Europe both Sevilla, Spain and Cantania, Italy are near to 37 degrees latitude.

It is important to note that a solar module output will be continuously variable during the year and even during the day. In wintertime it will output less power than its rating and in summertime it will frequently output more power. These electrical ratings are for guidance only and it is where many new photovoltaic designers make mistakes in thinking that the module will never output more power than its rating. It is important that you understand that these solar photovoltaic modules can output far more power than their label states. It can be over fifty percent more and this will need to factored into the system design.

Temperature Adjustments

Solar modules are affected by temperature, both hot and cold, and adjustments to the module ratings needs to be made for the operating temperature outside of 25 degrees Celsius. It is important when designing a system that the historical temperature minimum and maximum values are known for the area where the system is to be installed and these adjustments are factored into the design.

Open Circuit Voltage (Voc)

The open circuit voltage rating is how much voltage the module will put out with no load attached. This is an important value in order to design a system. This is the voltage to use when selecting your components and it must be adjusted for the historical minimum and maximum temperatures for the area. If more than one module is connected in series then multiply this temperature adjusted voltage by the number of modules in series to get the total maximum DC voltage of the system.

DC Maximum Power Point

The DC maximum power point is a simple concept. Power is a function of both voltage and current. The maximum power point is obtained when the current and voltage from the module when multiplied together give the maximum power figure. These values will change constantly during the day with the weather conditions. Voltage will remain relatively constant, but current will vary a lot with irradiance. The DC to AC inverter system constantly monitors the power from the solar photovoltaic DC system and automatically keeps the inverter system working at the maximum power value for the given conditions.

DC Maximum Power Point Voltage (Vmpp)

The DC maximum power point voltage (Vmpp) is the operating voltage of the solar module under load. Again this value will change with temperature and irradiance, but should only vary by about twenty percent of the STC rating during the day time.

DC Short Circuit Current (Isc)

The DC short circuit current value is the maximum current that the module will output at standard test conditions if the positive and negative terminals were connected (shorted) together. It is important to note that this value will vary a lot dependent on weather conditions and can be over fifty percent larger during summertime.

DC Maximum Power Point Current (Impp)

The DC maximum power point current is the amperage that the solar module will output at standard test conditions in normal operation. It is important to note that this value will vary a lot dependent on weather conditions and can be over fifty percent larger during summertime.

DC Rated System Voltage

This is a very important design value. It is a rating of how many modules can be safely connected together in series, this is called a solar photovoltaic module string. This system voltage should never be exceeded when adjusting for the minimum and maximum temperatures of the area that the system is being installed. This value basically limits the number of modules that can be connected in series in the system.

3. Photovoltaics and Weather

The performance of any solar photovoltaic system is dependent on the weather. The main factors that affect the system performance are clouds, irradiance, temperature, shade, latitude and how dirty the solar modules are. Let's now explore the effects of the weather in more detail:

Irradiance

Irradiance is a measure of how much sunlight the solar module is receiving. It is given in watts per meter squared or W/m^2. Standard Test Conditions (STC) uses a value of $1,000W/m^2$. This value can range from $0W/m^2$ at night through to over $1,500W/m^2$ during a day interspersed with large fluffy clouds. This value of $1,500W/m^2$ is larger than what you would receive in space. The reason why we can get greater values at ground level is due to what is known as the "cloud effect". Normally the sunlight is traveling in a straight line from the sun to our solar module with some atmospheric scattering. However, when clouds are present they can also reflect and can act like lenses to send some extra sunlight onto the solar modules. This extra light is converted into extra energy and this is seen largely as an increase in power from the system. This effect can be a few minutes long in duration when it occurs.

The diagram on the next page demonstrates the "cloud effect".

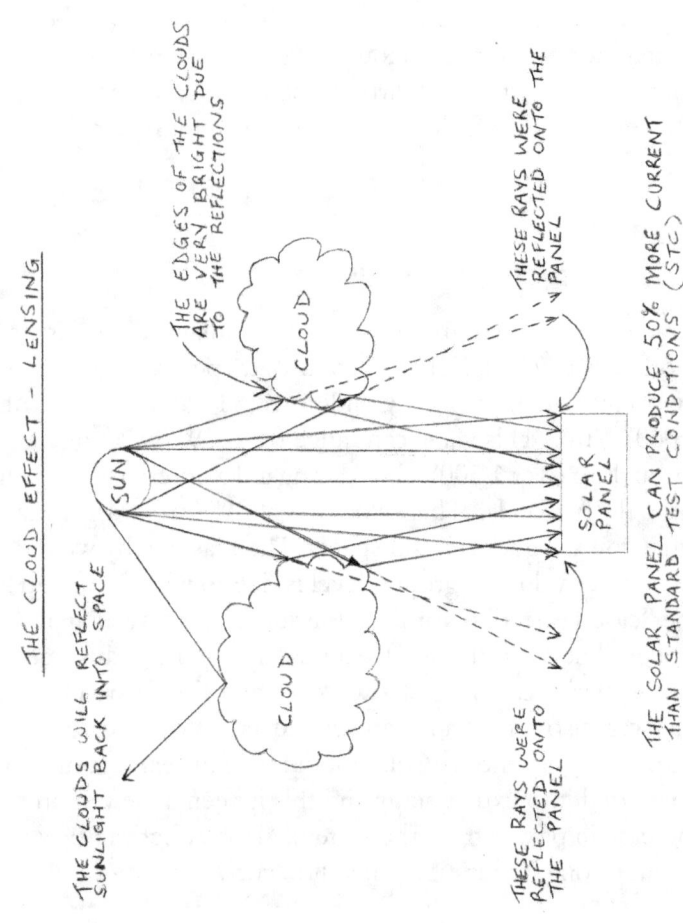

THE CLOUD EFFECT - LENSING

THE EDGES OF THE CLOUDS ARE VERY BRIGHT DUE TO THE REFLECTIONS

THESE RAYS WERE REFLECTED ONTO THE PANEL

THE CLOUDS WILL REFLECT SUNLIGHT BACK INTO SPACE

SUN

CLOUD

CLOUD

SOLAR PANEL

THESE RAYS WERE REFLECTED ONTO THE PANEL

THE SOLAR PANEL CAN PRODUCE 50% MORE CURRENT THAN STANDARD TEST CONDITIONS (STC)

Other effects on irradiance are the snow effect, the lake/ocean effect, the building effect and albedo. Snow cover, water, glass covered buildings, reflective painted buildings and roofs, and the albedo of the area surrounding the solar photovoltaic modules can reflect extra sunlight onto the solar power system. If you are installing a system in an area that has any of these, it is important to account for it. Each effect can produce an increase in power output. If you find yourself having to wear sunglasses in the solar photovoltaic system location for your eyes to be comfortable, then you probably have light reflections taking place.

In wintertime generally the system will operate at below the standard test conditions values and in summertime it will generally exceed these values. During the design phase of the system you will need to assess where the greatest power need is and perhaps increase the size of the system accordingly if it is in wintertime.

Air Mass

Air mass is a measurement of the the amount of atmosphere that the sunlight has to pass through to get to the ground. It varies with the seasons and also the location on the earth. Within the tropics, air mass will reach its maximum power value of 1 during summertime. Air mass 1 corresponds to the sun being directly overhead, air mass increases as the sun moves from directly overhead down to the horizon.

All USA solar modules are rated for air mass 1.5 which corresponds to a central USA location in Spring and Fall. When in a southerly location you will approach air mass 1 which will increase power output by about 13% from STC in the USA.

Locations that are at or near air mass 1 in Summer time in the USA are all Hawaiian islands, Florida and Texas. Approximately half of the continental USA is located between air mass 1 and air mass 1.5 in Spring and Fall. If you are working on systems that are located in these Southern USA states, you will get more power out of these systems due to a decreased air mass. In summertime the air mass will move closer to 1 in the continental USA.

Clouds

Clouds come in many forms. An important question is how do clouds affect irradiance on solar power photovoltaic power systems? The list below will help with understanding the effects of clouds on irradiance at air mass 1 (within the tropics in summer time):

- Clear, sunny skies will give approximately 1,130W/m^2. The transmission characteristics of the atmosphere will vary in clear skies, sometimes being relatively transparent and other times being more opaque and this affects irradiance values. Air quality is a major factor for the transmission of sunlight through the atmosphere. Particulate matter in the atmosphere will reduce the transmission level.

- Thin cirrus will give approximately 1,000W/m^2. Thin cirrus will give even and relatively stable irradiance levels due to scattering of the light.

- Thick cirrus will give approximately 750W/m^2.

- Thin clouds will give about 500W/m^2.

- Thick clouds will give about 250W/m^2. No shadows on the ground will be present

- Thick clouds with a visibly dark sky will give about 100W/m^2. No shadows on the ground will be present. You will not be able to see the location of the sun in the sky.

- Tall broken clouds will give surges of about 1,500W/m^2 and reductions to about 100W/m^2 of irradiance due to the cloud effect. The rate and length of time for these surges and reductions is dependent on the speed of the clouds passing in front of the sun.

Temperature

Temperature will affect the system to a much lesser extent than irradiance. The cooler the system is below 25 degrees Celcius, the more power it will produce. Correspondingly, the hotter the system is above 25 degrees Celcius, the less power it will produce. Temperature can affect solar photovoltaic systems power output by about twenty percent.

Shade

It is undesirable to shade solar modules as it can significantly affect the performance of the system. When studying the location of where to install a system, always factor in the surroundings for shading effects. Avoid shading with solar photovoltaic power systems.

Wind

Wind will provide cooling to the photovoltaic modules and it is an aid to power production. A breezy location will provide improved performance from the system. When mounting solar modules onto racking, it is good to allow spaces between the solar modules in order to aid with cooling airflow around the modules and also to reduce wind resistance. When choosing solar modules and mounting systems, it is important to ensure that they are rated for the wind speed of the area that you are installing them into.

Altitude

A higher altitude location will improve the amount of irradiance that the system will receive, due to less scattering and absorption of the sunlight by the atmosphere. It also acts as a natural cooler of the system which further improves system performance. Generally a high altitude location will have a higher percentage of clearer skies during a year which will give a higher energy yield from the system.

Snow and Ice

Snow and ice should not affect a solar module, other than obscuring its view of the sun. Tracking systems can be affected by this and in some snowy locations it is advisable to park the solar system facing South during these periods. The reflection from the snow will increase the power from the system in Winter time.

Hail

Hail can break solar modules, so it is important to know type of hail that your area can receive. If you get large golf ball size hail, you may not want to install glass solar modules. Solar modules are tested for hail and pass the tests even if the glass module breaks. The test just ensures that the modules remains intact when broken. Glass solar modules are hard to break and normal sized hail should have no effect.

Dirt

Clean solar modules are the desirable configuration for a system. However, dust and dirt will get onto the surface of the modules and will degrade performance by up to 10% on average. Cleaning the modules is very much a function of the location where they are installed and also how dirty they are. Most people will clean on an as needed basis, generally when they are visually very dirty. Always follow the manufacturers instructions for cleaning your particular modules and remember that solar modules are operating with electricity flowing in them when exposed to light. Night time cleaning is recommended for safety.

Lightning

Lightning can affect solar modules, especially on large systems that cover fields. Good equipment grounding is the way to deal with this threat. A low resistance ground will generally dissipate lightning away from a solar module that is struck by lightning. Generally, the damage should be limited to only the solar module that was struck. If a cable is struck, then lightning surge arrestors can limit the damage in the system. These are generally installed in the inverter and on larger systems, in combiner and re-combiner boxes. Install lightning protection as recommended by the manufacturers of the products used in the installation.

Seasons

We have four distinct seasons of Winter, Spring, Summer and Fall. We can word this another way as Winter Solstice (December 21), Spring Equinox (March 20), Summer Solstice (June 21) and Autumn Equinox (September 22). What does this mean to a solar power system?

- The length of the day
- The angle of the sun (air mass)
- Heating and cooling
- Rain

Winter solstice is the shortest day of the year and summer solstice is the longest day of the year. Spring and fall equinoxes are when day time is the same length of time as night time.

Regarding the angle of the sun in the USA, Winter Solstice is when the sun is at the lowest in the sky, or 23.5 degrees below the equator and Summer Solstice is when it is 23.5 degrees above the equator. Spring and Fall equinoxes are when the sun is directly overhead at solar noon at the equator.

For our solar power system, this means that we will produce our largest voltage in wintertime when it is the coldest and we will produce our largest current when it is summertime with peak irradiance.

The changing seasons will affect rainfall and in dry seasons you may want to schedule cleaning to keep the modules in good performance. Rain generally helps to keep the module clean naturally.

There are a number of things to consider with the seasons:

- Spring & Autumn
    - The system will be operating close to standard test conditions (STC) and measured values should be close to that on the solar module label.
    - This is the most favorable time for outdoor working.
- Summer time
    - The system will be hot and the DC voltage will be lower than normal.
    - The system DC electrical current will be at the highest value for the year.
    - Ambient temperatures will be high.
    - Heat and dehydration may be a problem for working on the system.
- Wintertime

- The system DC voltage will be at the highest value for the year
- It may be too cold to work on the system
- Frost, ice and snow may be an issue for performing maintenance.
- Ambient temperatures will be low.

Due Diligence

It is important when designing, operating and maintaining a solar power generation system that you are aware of the annual climatic conditions to expect. Amongst the data that you should have is:

- Historic annual minimum temperature
- Historic annual maximum temperature
- Historic annual maximum wind speed
- Historic annual snow fall depth
- Historic annual hail size
- Historic annual peak irradiance
- Historic monthly irradiance

With these values you will be able to make educated engineering decisions regarding the selection of your system.

4. System Selection

System selection is the most complicated stage of solar photovoltaic system designs and will determine how well your system works. When done well, your system will give years of trouble free operation.

Solar Modules

Solar modules typically represent the largest investment in the system. Traditionally silicon solar modules have been used, but now many new types of technologies are emerging such as thin film and so on. All solar modules are tested to the Underwriters Laboratory (UL) 1703 standard in the USA.

Silicon is the best understood, the most efficient and has been around for many decades. The newer film type solar modules are less efficient and cheaper to purchase. Unfortunately more thin film modules are needed to generate the same power as silicon, and this increases the system physical size and associated support systems such as cabling, racking, installation costs and so on.

Generally any decision on which technology to use is driven by market rates for each type of technology, aesthetics and personal preference. Solar modules are a commodity and their prices can fluctuate rapidly.

Mounting System

There are three ways to mount your modules:

- Fixed tilt
- Single axis tracker
- Dual axis tracker

All have their pros and cons.

The fixed tilt system is the most common and is widespread. The solar modules are either mounted to a roof, building or are ground mounted in a fixed position inclined to face south at a tilt angle matching the the latitude. Some systems allow you to adjust the tilt angle of the modules for the season, but it appears that most people prefer the low maintenance option of mounting the modules into a fixed position for the entire year. The fixed tilt system is the most reliable configuration and also the lowest cost. The downside is it has the lowest annual energy output of the mounting systems.

The single axis tracker works well. The solar modules are mounted on a rotating North-South axis which allows them to track from East to West during the day. There are two types of single axis trackers generally available. The first has the North-South axis mounted horizontal and the modules can track in the East to West direction. This system works well in or near to the tropics where the sun can be almost directly overhead. The second has the North-South axis inclined to match the latitude and this enables the solar modules to face the sun in spring and fall. This system works better as you move further away from the tropics. A single axis tracker can increase annual energy output by about 25% when compared to a fixed tilt system. The single axis tracker does not cost much more than a fixed tilt system and the extra expense is generally offset by the extra annual energy yield of the system. The downside is that the tracking mechanism does need regular maintenance and occasionally will break down.

The dual axis tracker has the modules tracking the sun from sunrise to sunset, keeping the solar modules in the optimal position for maximum power generation. A dual axis tracker can increase annual energy output by about 40% when compared to a fixed tilt system. The downside to a dual axis tracker is that it requires a lot of space, can be very tall, has a complicated control system, they are expensive and they are the highest maintenance system.

The diagram on the next page shows the differences for each tracker system at noon with the seasons.

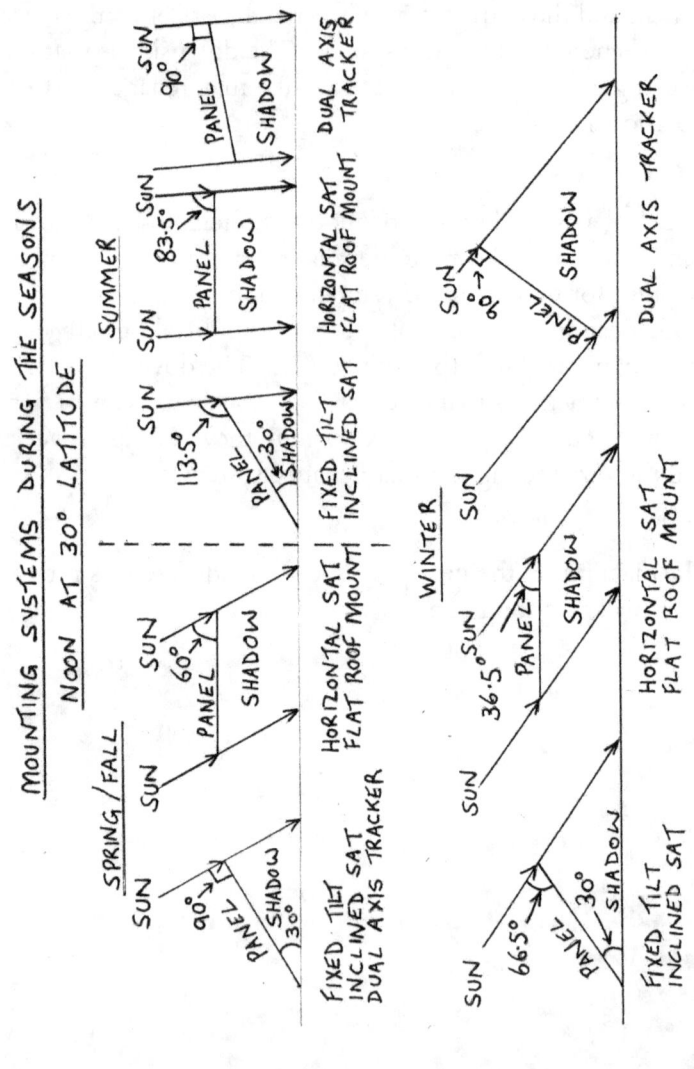

Cabling

For all outdoor cabling exposed to the sun, solar rated cable should be used due to its resistance to ultraviolet radiation. Once inside a building, conduit or underground this requirement need not apply.

On ground mounted installations it is important to remember that animals will be able to access the cables and equipment, so you will need be familiar with your local wildlife in the area. Where possible, enclose all accessible cabling with protective covers.

Conduit or ducting is recommended for all underground runs as it can be easily rewired to a larger size if you find that you system is generating more power than designed for or if a cable goes faulty.

Inverters

Inverters come in many different types and sizes. This book is dedicated to grid tie inverters and this is what we will consider. All residential and commercial solar photovoltaic grid tie inverters have to comply with UL1741 in the USA. This ensures that all grid tie inverters meet the requirements of the utility authority. Some of these requirements are below:

- Disconnect from grid during power cuts

- Protection against faults

- Good power quality output

Solar Design Hints and Tips

All DC Grid Connected Solar Photovoltaic Systems Need

- Equipment Ground.
- DC Disconnect.
- Inverter.
- AC Disconnect.

As They Get Larger They Also Need

- Combiner box.
- Recombiner boxes.
- Transformers and switchgear.
- Distribution and transmission systems.

Important Guidelines for Tracking Systems

- Try to put no more than one tracking system on one inverter.
- Install blocking diodes in each string to prevent reverse current flowing on multiple tracking systems installed on one inverter.

DC component selection

- Always make sure that DC rated components are used.

- Components must exceed specifications for maximum voltage.

- Components must exceed specifications for maximum continuous current.

- All outdoor cables must be solar rated.

- All outdoor connectors must be solar rated.

- De-rate equipment, cables and fuses for installed conditions such as:

  - High ambient enclosure temperatures.

  - Burial.

  - Enclosed conduits and ducts.

  - Heating effects.

  - Cooling effects.

  - High power cycling.

- Cables, breakers and fuses:

  - At least 156% larger than the short circuit solar photovoltaic current.

  - Increase current as needed for reflections.

  - Never exceed the interrupt ratings of the fuses and breakers.

- Use metal enclosures for effective heat dissipation from the electrical equipment.

AC Component Selection

- Always make sure that AC rated components are used.

- Components must exceed system specifications for maximum voltage.

- Components must exceed system specifications for maximum current.

- All outdoor cables must be solar rated.

- All outdoor connectors must be solar rated.

- De-rate equipment for installed conditions such as:

    - High ambient enclosure temperatures.

    - Burial.

    - Enclosed conduits and ducts.

    - Heating effects.

    - Cooling effects.

    - High power cycling.

- Cables, breakers and fuses:

    - At least 125% larger than the maximum AC current.

    - Apply de-rating for highest expected fuse ambient temperature per fuse manufacturers de-rating tables.

    - Never exceed the interrupt ratings of the fuses and breakers.

    - Use metal enclosures for effective heat dissipation.

Enclosure Mounting

- Exterior equipment locations should always be in the shade.

Inverters

- More inverters improves maximum power point tracking on the system

- Keep inverter power down to about 250 kW AC maximum per single inverter.

- Try to keep below 100 strings per inverter.

- If possible, mount inverters in shaded locations, consider constructing a shade canopy if needed.

- If inside a building or structure, ensure that the indoor ambient temperature can never exceed the inverter ambient temperature ratings.

- Use proven inverter technology for your system size.

Manufacturer Instructions

- Always design your system in accordance with the manufacturers installation manuals

- Follow the maintenance schedules in order to maintain warranty coverage

Codes

- Always design to local photovoltaic electrical codes.
- Always design to local building codes
- Consult with qualified engineers
- Have qualified engineers approve designs
- When in doubt, engineer on the side of safety.

Labeling

- Make sure all equipment is labeled in accordance with the local codes and the manufacturers instructions.

Safety

- Make sure all system and equipment grounding is installed correctly
- Make sure all covers are installed
- Make sure DC polarity of cabling is correct
- Make sure AC phase rotation is correct for three phase systems.
- Follow the local safety codes

System Inefficiencies

The system inefficiencies comprise of the following component de-rate factors:

- PV module nameplate DC rating 0.80 - 1.05
- Inverter and Transformer 0.88 – 0.98
- Mismatch 0.97 - 0.995
- Diodes and connections 0.99 – 0.997
- DC wiring 0.97 - 0.99
- AC wiring 0.98 – 0.993
- Soiling 0.30 - 0.995
- System availability 0.00 - 0.995
- Shading 0.00 – 1.00
- Sun-tracking 0.95 – 1.00
- Age 0.70 - 1.00

A well built solar photovoltaic system should be in the vicinity of 77% to 85% overall efficiency conversion from DC to AC at the point of interconnection.

System Size

There are two ways that are traditionally used to quote system sizes

- Installed solar module DC power at STC

— Expected AC output at system interconnection at STC

It is important when designing a system that you quote both values accurately to the customer.

5. Grid Interconnection Requirements

Grid interconnection is different all over the world and it will depend on who's grid you are connecting into. Each utility will have its own requirements for grid interconnection and you will need to contact them to obtain their documentation on this.

In general they are looking for the following attributes:

- Address of installed system
- System generation capacity
- Power factor of system generation
- Inverter manufacturer(s) and model number(s)
- Sufficient interrupt ratings on breakers and fuses
- For residential and commercial installations only, confirmation of anti-islanding feature.
- Certification of country standards of manufacture of equipment.

It is important as a utility customer that you use equipment that will disconnect upon the grid power failing. This is to ensure that the grid is not back fed and energized from your equipment. This represents a significant hazard to utility line workers and many have been exposed to electrical hazards caused by this effect in the past. UL 1743 listed inverter equipment meets this requirement.

For a large utility scale system, you will want the utility to supply the following details:

- Interconnection fault current
- Interconnection voltage
- Transformer configuration
- Relay settings

With these values you will able able to design your utility interconnection.

6. Residential Systems

Residential design for the purposes of this book is systems up to 10 kWp DC at STC. Most readers of this book will be working with this size of system. Residential systems represent the starting point for our learning and underpin the larger system designs. To understand solar photovoltaic systems, this is your starting point.

Residential systems are relatively easy to design electrically and have very few solar strings, sometimes they may only have one string. Something to consider with a residential system is the size of the solar modules that you will use. Since most solar systems on a residential system will be mounted to the homeowners roof, it is wise to select a solar module that is easily handled when up on the roof. Sometimes bigger is not better when installing roof mounted systems.

Most solar photovoltaic systems will be mounted onto roofs. Certain roof types are unsuitable for solar photovoltaics and you will need to assess this. You will also need to make an assessment to make sure that the roof is strong enough to mount the solar power system to. A structural engineer can perform this assessment, if needed. A good starting point with a residential system is to discuss the requirements for a solar photovoltaic system with the local planning department to see what they require.

Once confirmed that you can mount a solar photovoltaic system to the roof, you will need to identify where the roof supports or trusses run. You will want to attach your mountings to this location. If it is tiled, a roofer can help you with removing the tiles and attaching your chosen mounting to the roof. Sometimes you will be able to drill the tile and refit it, other times you will want the roofer to flash the mounting with lead to waterproof the location in the absence of the tile being refitted.

A sample residential system is shown on the next page and illustrates the concepts of residential solar photovoltaics. Note that the point of interconnection is at the electrical meter due to the fuse board not being rated for solar photovoltaic (bi-directional power flow) use. We know this because the fuse board had "Line" and "Load" marked on its terminals. Bi-directional fuse boards have no "Line" and "Load" markings on their terminals.

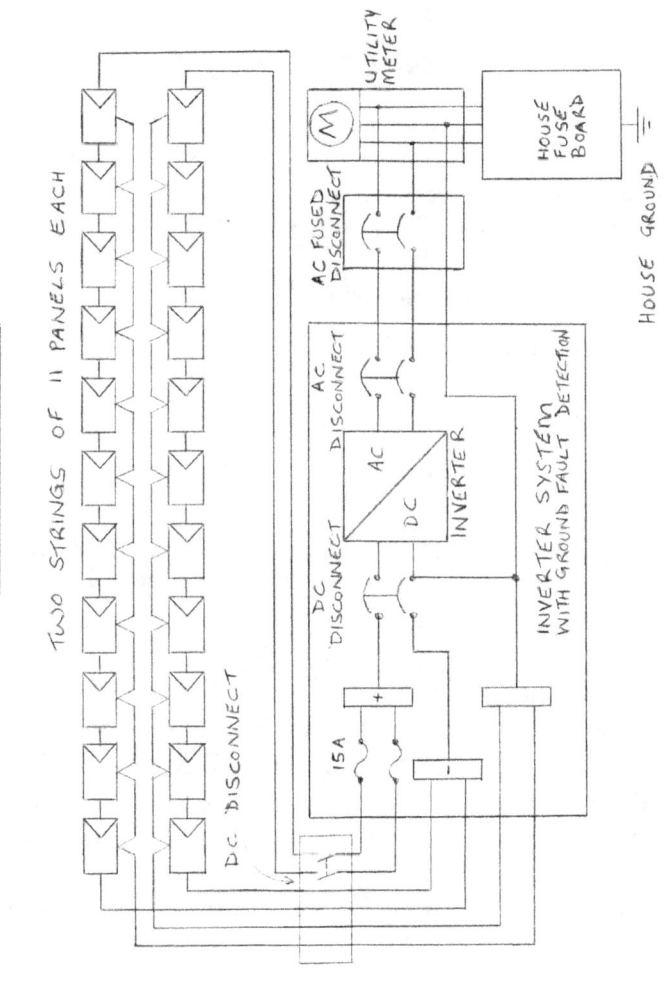

7. Commercial Systems

Commercial systems for the purposes of this book is systems from 10 kWp DC to 1 MWp DC at STC.

New ideas in this section are

- MPPT tracking performance
- Combiner boxes

MPPT Tracking Performance

Maximum power point tracking (MPPT) is a feature that is built into the inverter and the it keeps the DC system functioning at the maximum possible power output automatically. Maximum power point tracking (MPPT) is at its most efficient on a single solar module string. As you increase your system size and add more strings to it, it can reduce the performance of MPPT. The MPPT tracking will only be as good as the worst performing strings connected to it. As such it is important to limit the number of strings connected to an inverter. I would suggest that no more than one hundred strings be connected to an inverter for this reason.

Combiner boxes

These are needed to connect our strings together so that they can be combined into a larger cable. They generally come in 6, 12, 18 and 24 string capacities. More advanced models have the capacity to monitor each string connected to it using a computer network and the associated software. Each string will have a fuse or circuit breaker for it.

The diagram on the following page illustrates the concepts of commercial electrical design for a single 250 kWp DC inverter system.

250 KWp DC SYSTEM
ONE LINE DIAGRAM

8. Utility Systems

Utility design for the purposes of this book is systems over 1 MWp DC.

A new set of guidelines comes into play with these systems:

- Power Factor
- AC and DC Interrupt Current
- High Voltage Transmission
- Ride Through Capabilities
- Harmonics Requirements
- Large Power Swings
- Distributed or Centralized Generation
- Re-combiner Boxes
- National Electric Safety Code (NESC)
- 1,000 Volt DC Systems

Power factor

When generating power from an inverter it is generally set to unity. However, for a utility grid, this will cause problems with the power factor regulation on the grid. Feeding in a large amount of unity power factor energy into a grid that is operating at 0.9 lagging power factor will bring the overall power factor down to a lower level, such as 0.8 lagging. It is undesirable to operate the grid at this level and as such, control of the power factor being generated by the inverter(s) is desirable. Generally utility inverters can output their power factor between 0.9 leading and 0.9 lagging. The utility will inform you of the value to set the inverter to.

A nice feature of utility inverters is that the power factor can be controlled. This offers the ability of power factor correction for the grid. When designing a utility system, it is important to design in the ability to control power factor for the utility authority.

AC and DC Interrupt Currents

At the utility level, this becomes a major concern. All fuses and circuit breakers have interrupt ratings. It is important when working with interrupt currents that you ensure in the DC circuit that you are using the DC interrupt current ratings and in the AC circuit that you are using the AC interrupt current ratings. Generally, interrupt currents are in the range of tens of thousands of amps.

If you are using a large inverter, then it may have an equally large capacitor inside of it. It is important to find out the short circuit fault current for the inverter DC bus as you will need this to verify the interrupt current of your DC design. Most inverters do not allow back feed into the DC system under fault conditions, so this may not be an issue on most systems. My recommendation is to use a DC design of 250 kilowatts STC or below to keep the DC fault currents to a manageable level.

You will need to ensure that you do not exceed the interrupt current of the fuses and circuit breakers used in the system. Sources of current that will need to be considered in the DC circuit are:

- Solar module maximum circuit current multiplied by the number of solar module strings connected to the inverter.

- Cable capacitance

- Inverter capacitance

In a inverter system of over a megawatt with several hundred strings spread over acres of land, these interrupt currents could be very large. Be careful when designing such large DC circuits and pay close attention to the DC interrupt current.

Long cable runs that are present on the DC circuits will generate capacitance effects. The cable manufacturers data sheet will specify the capacitance per unit length of the cabling and the total capacitance can be calculated for the cabling.

In the AC circuit, the largest fault current will generally be that from the grid. Before selecting your AC equipment, you will need to establish this value from the utility and calculate the fault currents for your particular design.

These are many sources of fault currents in the AC circuits:

- The grid

- The solar power inverter systems

- Cable capacitance
- Power factor control capacitance
- Inductance

Make sure that any AC fault analysis includes all sources of power feeding into the fault.

Always calculate the fault currents for the DC and AC circuits and ensure all fuses and breakers can interrupt these currents.

High Voltage Transmission

Transmission for the utility grid can be at over one million volts. The previous designs that have been discussed operate at low voltages. Transforming the voltage up from this to medium and high voltage is straight forward.

You will need to know the following items to design the transmission interconnection:

- Interconnection voltage.
- Star or delta feed into the transmission network.
- Interconnection short circuit fault current.
- Transformer impedance.
- Harmonics requirements.
- Protection relay system required by the utility.
- Protection relay settings required by the utility.

With these, you will be able to design the solar photovoltaic system interconnection. It is important when designing the interconnection that you use electrical switchgear that is capable of being back fed in your design.

Ride Through Capability

Utilities want to generate power and if the grid voltage dips, they want their power generations systems to stay on line and "ride through" the dip. This is a different concept than previously discussed and is unique to the utilities.

It is important when building a system for a utility that you select a utility grade inverter that has this feature. The utility wants to keep power feeding into the electrical grid and wants the grid to stay on line. Generally the ride through feature can be set to keep feeding into the grid for a specified time before the inverter will shutdown. Check with the utility what their requirements are for this feature.

Harmonics Requirements

Most utility inverters generate power at below 3% harmonic distortion. This is generally acceptable by the utilities and multiple inverter systems can keep the harmonic distortion below this level. When selecting your inverter system, it is always good to get confirmation from the manufacturer that a large, multiple inverter system can keep within this tolerance.

Large Power Swings

Any solar photovoltaic system will suffer from large power swings that are proportional to irradiance. Solar cells directly convert irradiance into energy in real time. The main thing that can significantly effect the power output quickly are clouds. Clouds can increase the power as well as decrease the power output. Large, dense, fluffy broken clouds will reflect light from the sun and increase the power output from the solar modules that are in direct sunlight. As the cloud starts to approach the sun, lensing will occur that will focus more sunlight into the solar module. This surge in power can be a ramping up of normal power levels as higher than 25% increase for a few minutes. As the cloud progressively covers the sun, the power will start to drop off significantly with drops of over 90% from normal power levels. On a broken cloud day, these large power swings can occur within several seconds if the cloud is fast moving or over several minutes if they are moving slowly.

As you can imagine, these large power swings cause havoc with the grid and can look like faults to the grid control system. On a large power generation system, it is important to build into the grid control system feedback from the solar power system control system so that the grid can discern between a genuine fault and clouds passing over the solar power generation system.

The grid needs to be able to tolerate these large swings and alternate conventional power generation needs to be able to rapidly react to the power surges that the solar power system will generate during broken cloud days.

Distributed or Centralized Generation

Distributed power generation refers to when the solar power generation system is broken into many small parts and spread over a wide area. Residential installations and commercial installations would fall into this category. Distributed power generation is less prone to broken cloud effects due to the solar modules being spread over a large area. On a broken cloud day, some solar power systems will be in the sun while others will be in the shade. It creates an averaging of the broken cloud effect on the grid and is much more desirable.

Centralized power generation is when a large solar power generation system is installed in one location. Utility scale installations would be in this category. There are advantages regarding maintenance costs to centralize the utility solar power plant, but these seem to be negated by the cloud effects on the solar power system output. Centralized power generation works well when complimented by another source of fast response power generation, such as a gas turbine power generation plant. Thus the solar power generation system is used to offset the fuel consumption of the conventional power generation plant.

Recombiner boxes

Recombiner boxes are used to combine the outputs of two or more combiner boxes into a larger cable. They generally will have two to four large circuit breakers inside them, depending on the amount of combiner box circuits being combined.

National Electric Safety Code (NESC)

Now that we are on the utility side of the meter, a new standard applies. The National Electric Safety Code (NESC) covers the utilities. One thing that you will notice is that there is no mention of solar photovoltaics in the National Electric Safety Code. Due to this, we obtain our solar photovoltaics design equations from the National Electric Code (NEC). The National Electric Safety Code (NESC) and the National Electric Code (NEC) should be read in conjunction with each other. You will find on utility scale installations that you will refer to the two code books frequently.

1,000 Volt DC Systems

There has been a move towards 1,000 volt DC systems in the USA. 1,000 Volt systems are widespread in Europe. The advantage is that the efficiency of the DC circuit is increased due to lower power losses in the circuit. Cabling and equipment costs are reduced as less is used. I personally do not recommend these systems as a solar photovoltaic systems need maintenance at least annually and this high voltage increases the electrocution risks for the workers performing this maintenance. National Electric Code section 490.2 defines high voltage as over 600 volts. As such the provisions in National Electric Code section 490 as well as the National Electric Safety Code (NESC) should be followed.

A sample design showing the concepts for a 10MWp DC solar photovoltaic utility system is shown on the next pages.

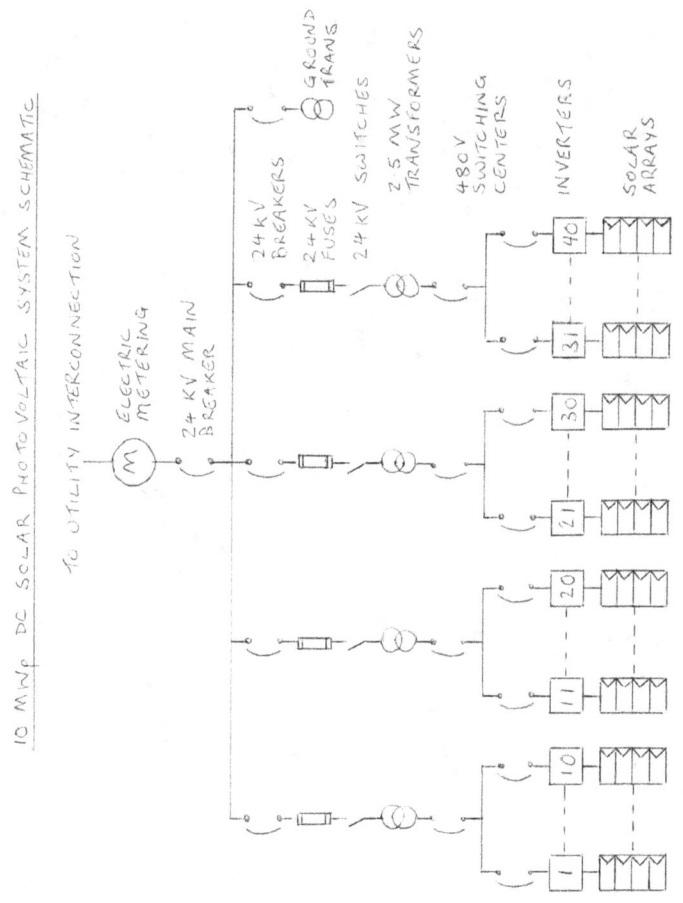

10 MWp DC SOLAR PHOTOVOLTAIC SYSTEM SCHEMATIC

To UTILITY INTERCONNECTION

ELECTRIC METERING

24 KV MAIN BREAKER

24 KV BREAKERS

24 KV FUSES

24 KV SWITCHES

2.5 MW TRANSFORMERS

480V SWITCHING CENTERS

INVERTERS

SOLAR ARRAYS

GROUND TRANS

40
31
30
21
20
11
10
1

10 MVA DISTRIBUTION

9. System Sizing

Sizing your system is a complicated equation of your energy demands and the area that you install your system.

"Net zero" is a common term in the solar industry. This means that your solar photovoltaic system is sized to match your annual consumption of electricity from the grid. In other words you use no electricity from the grid during the year when averaged out over the year. If you are looking for a net zero system, then the first step is to find out how much electricity you use during the year from your electricity bills.

Divide the annual energy consumption by 365 days to get the daily figure.

The next step is to look at the annual solar radiation charts for your area. These can be obtained at the National Renewable Energy Laboratories website at www.nrel.gov. The NREL USA chart for annual average solar data for photovoltaic systems is shown on the next page and was obtained from http://www.nrel.gov/gis/solar.html. I would recommend that you take a look at it in color at the website and download it. You can find out an estimate of how many hours of average daily solar radiation that your area will get from these charts.

Solar photovoltaic system size will be the daily energy consumption divided by the daily solar radiation.

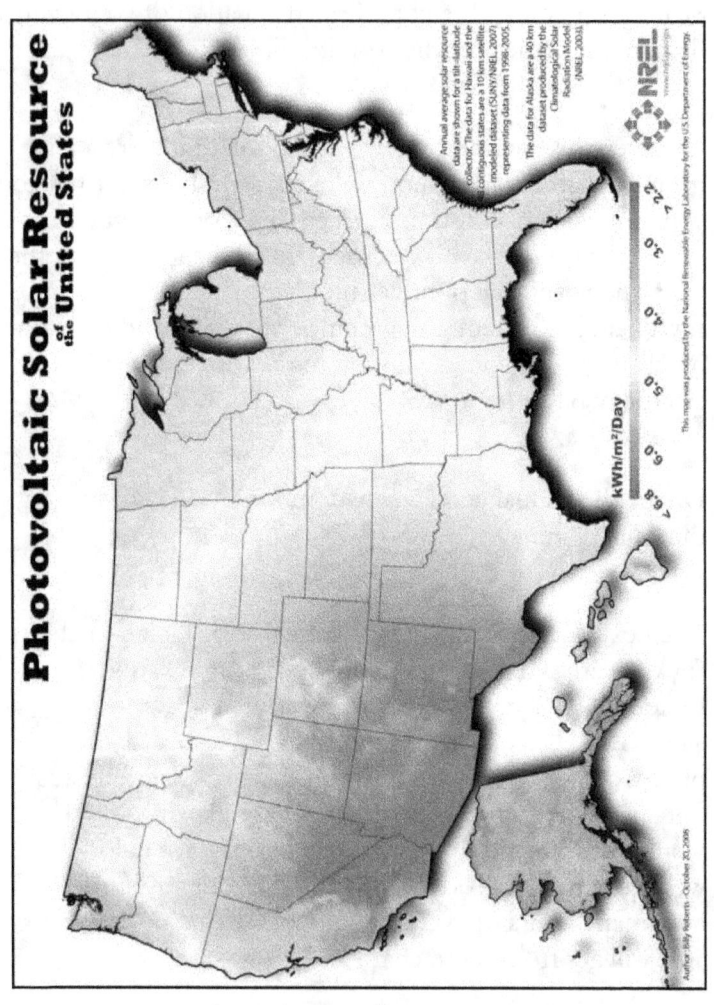

Using the commonly used figure of 0.77 DC to AC derating for the solar power system, adjust the system size figure for the photovoltaic system efficiency losses.

You can adjust for aging effects using a 1% per year of loss of efficiency.

For example:

This particular sample household uses 7,300 kilowatt hours (kWh) of electrical energy per year.

The system that will be installed will be a South facing fixed tilt system that is inclined to match the latitude angle. There is no solar photovoltaic module shading during the day.

Divide 7,300 kWh per year by the number of days per year

= 7,300 / 365

= 20 kWh per day

Looking at the solar radiation charts, this area gets an an annual average of 5 kilowatt hours per square meter per day.

Solar radiation = 5 kWh/m²/Day

We now divide the daily household power requirements by the number of hours of solar energy for this area.

20 kWh / 5

= 4 kW AC solar photovoltaic system size

We need to adjust this for the solar photovoltaic system DC to AC derating of 0.77 to get the correct DC system size.

= 4 kW / 0.77

= 5.19 kWp DC solar photovoltaic system size

Since solar photovoltaic power systems do suffer from aging effects by about 1% per year reduction in power from new, we will factor this in at a 10% loss after 10 years of service and increase the system size accordingly to compensate.

5.19kW / 0.9

5.77 kWp DC

So our system should produce more energy when new and after ten years will hit the net zero figure as it ages.

It is always good practice to oversize solar power systems so that the owner is not disappointed by poor performance. There is nothing worse than a system that does not meet its minimum performance expectations, which in this case is a net zero system.

A further adjustment is required for tracking systems:

A single axis tracker that is inclined at the latitude angle can produce an increase of energy production by about a factor of 1.24. We will reduce the DC system in size accordingly which gives:

= 5.77 kWp DC / 1.24

= 4.65 kWp DC single axis tracker system size

A dual axis tracker can increase the power production by about a factor of 1.3. We will reduce the system size accordingly:

= 5.77 kWp DC / 1.3

= 4.44 kWp DC dual axis tracker system size

You must remember that solar photovoltaics is more of an art than a science and these figures are approximations. It is common for some systems to over perform and others to under perform. It really just depends on your local conditions.

The safest way to quote solar photovoltaic power generation systems is installed DC capacity and installed AC capacity after conversion losses to the point of interconnection with the grid at STC. Annual power production predictions can be unreliable at times and should not be relied on.

Most companies when building a large power plant will build in expansion to the design at the start of the project as an insurance policy in case the system under performs in its location. Generally, a 10% expansion buffer is sensible to use for this purpose. It is wise to oversize the system to exceed the desired annual power production figure and to build it bigger if it under performs once in operation.

The above calculations are based on averaged data from NREL. For a large solar site, they should not be a substitute for a on site study of climatic and system annual performance for at least one year in duration prior to design and construction of a project. It is good to obtain several years of data if possible for a solar photovoltaic site prior to development.

The NREL website has a huge amount of historical solar radiation information on it and it is well worth taking the time to check out their latest information. The NREL data should be the starting point for planning a large solar photovoltaic project.

Sizing for power production for a certain month of the year is occasionally desired. For example, a common request is that a utility will want the solar photovoltaic power plant to cover the peak summertime load of the air conditioning loads on their system. Usually this type of load peaks in July.

To do this we look at the monthly solar radiation NREL charts and this is shown on the next page for July. It was downloaded from http://www.nrel.gov/gis/solar.html. I would recommend that you take a look at it in color at the website and download it. You can find out an estimate of how many hours of average daily solar radiation that your area will get over one month from these charts.

Using this chart, this area gets an average of 6 kilowatt hours (kWh) of solar radiation (called insolation) per square meter per day in July.

Solar radiation $= 6 \ kWh/m^2/Day$

We run the calculations, however, this time we factor in the temperature of the system as well as it will be hot. The peak ambient temperature for this area is 40 degrees Celcius and we will assume a average solar module surface temperature of 80 degrees for our calculations.

The utility wants to generate a daily average of 10 megawatts hours (MWh) of energy with the system in July.

Daily energy production $= 10 \ MWh \ per \ Day$

The system that will be installed will be a South facing fixed tilt system that is inclined to match the latitude angle. There is no solar photovoltaic module shading during the day.

We now divide the daily power requirements by the number of hours of solar energy for this area.

PV Solar Radiation
(Flat Plate, Facing South, Latitude Tilt)

July

Model estimates of monthly average daily total radiation using inputs derived from satellite and/or surface observations of cloud cover, aerosol optical depth, precipitable water vapor, albedo, atmospheric pressure and ozone resampled to a 40km resolution. See http://www.nrel.gov/gis/il_solar_pv.html documentation for more details.

kWh/m2/day
> 9.0
8.5 - 9.0
8.0 - 8.5
7.5 - 8.0
7.0 - 7.5
6.5 - 7.0
6.0 - 6.5
5.5 - 6.0
5.0 - 5.5
4.5 - 5.0
4.0 - 4.5
3.5 - 4.0
3.0 - 3.5
2.5 - 3.0
2.0 - 2.5
< 2.0

Produced by the Electric & Hydrogen
Technologies & Systems Center - May 2004

= 10MWh / 6

= 1.67 MW AC solar photovoltaic system size

We need to adjust this for the solar system DC to AC derate factor of 0.77 to get the correct DC system size.

= 1.67 MW / 0.77

= 2.16 MWp DC solar photovoltaic system size

Since solar power systems do suffer from aging effects by about 1% per year from new, we will factor this in for 10 years of service and increase the system size accordingly to compensate.

= 2.16 MWp / 0.9

= 2.4 MWp DC system size

Finally we apply the temperature correction for an 80 degrees Celsius solar module temperature which is -0.485% for power per degree on a silicon solar module, so we correspondingly increase the system size:

= System DC power + (system DC power x STC temperature difference x solar module power coefficient)

= 2.4 + (2.4 x 0.485% x 55)

= 3.04 MWp DC system size

So our system should produce more energy when new and after ten years it may hit the desired July 10 MWh figure as it ages.

It is always good practice to oversize solar power systems so that the owner is not disappointed by poor performance. There is nothing worse than a system that does not meet its minimum performance expectations.

A further adjustment is required for tracking systems:

A single axis tracker that is inclined at the latitude angle produces an increase of energy production by about a factor of 1.24. We will reduce the DC system in size accordingly which gives:

= 3.04 MWp DC / 1.24

= 2.45 MWp DC system size for inclined single axis tracker

A dual axis tracker can increase the daily power production by about a factor of 1.3. We will reduce the system size accordingly:

= 3.04 MWp DC / 1.3

= 2.34 MWp DC system size for a dual axis tracker

The utility may request that this system must produce a minimum amount of energy in summertime of 10MW on cloudy days. We would simply adjust our figures to account for this with an increase of system size to match the decrease in solar energy. For this example, the utility has requested 10 MW per day in irradiance conditions of 250 W/m². We have only 25% approximately of our sunlight, assuming July is a sunny month with clear skies, so we correspondingly increase the system size to account for this.

250W/m2 10MW Fixed tilt system:

= 3.04 MWp / 0.25

= 12.16 MWp DC system

250 W/m2 10MW inclined single axis tracker

= 2.45 MWp / 0.25

= 9.8 MWp system size

250W/m2 10MW Dual axis tracker

= 2.34 MW / 0.25

= 9.36 MWp DC system size

As you can see we have used a inverse approximation between irradiance and system size. As irradiance decreases our system size correspondingly needs to increase in size to keep delivering the same output power. This can be further refined by looking at the power output for the solar module technology selected for your system and using the actual power listed for $250W/m^2$ irradiation levels.

Now that we have a good understanding of how to perform solar photovoltaic system sizing calculations, let me introduce you to the easy way of doing this. Most solar photovoltaic system designers will use the PVWatt calculators at the NREL website http://www.nrel.gov/rredc/pvwatts/ to size their systems.

You must remember that solar photovoltaics is more of an art than a science and these figures are approximations. It is common for some systems to over perform and others to under perform. It really just depends on your local conditions.

10. Solar Modules

Solar modules typically represent the largest investment in the system. Traditionally silicon solar modules have been used, but now many new types of technologies are emerging such as thin film and so on. All solar modules should be tested to the Underwriters Laboratory (UL) 1703 standard or equivalent in the USA.

In the following chapters we will take a look at each technology and see the needs of each technology when building residential, commercial and utility systems with them. The main solar technologies in use currently are:

- Mono-crystalline silicon wafers (also called single crystal)

- Poly-crystalline silicon wafers (also called multi-crystalline)

- Amorphous silicon thin film

- Cadmium telluride thin film

- CIGS thin film

Silicon is the best understood, the most efficient and has been around for many decades. The newer film type solar modules are less efficient and cheaper to purchase. Unfortunately, currently more thin film modules are needed to generate the same power as mono and poly crystalline modules, and this increases the system physical size and associated support systems such as land purchase, cabling, racking, installation costs, operation and maintenance costs, and so on. Thin film uses less materials to produce it and it really is a film, usually about 1/50 the thickness of a silicon wafer solar cell.

Thin film is a relatively new technology in the solar photovoltaic field. Development of thin film systems was in process in the late nineties and became a commercial product in the last decade. Thin film benefited from a surge in silicon wafer costs and a lack of supply of silicon solar modules in the last decade. Thin film was widely installed during this time. Recently, this silicon supply problem has been rectified and silicon prices have fallen sharply and this has been hurting the thin film solar module manufacturers due to the superior efficiency of silicon wafers.

A further problem with thin film is that its longevity is still being assessed. Some thin film manufacturers offer a 25 year warranty but cannot demonstrate commercial product that is of that age. It will come with time, but financing companies are reluctant to finance thin film due to the higher risks that aging poses for these systems. In time, the various thin film solar modules are expected to be proven to have a similar reliability as silicon wafer solar modules.

Thin film systems are generally marketed as the solar panel of choice in cloudy locations due to their improved performance in low light conditions when compared to silicon. Thin film is less efficient than silicon and is currently between 7 and 10 percent in the commercial products. As such more solar panels are needed when compared to silicon systems to generate the same energy at STC.

Generally any decision on which technology to use is driven by market rates for each type of technology, aesthetics and personal preference. Solar modules are a commodity and their prices can fluctuate rapidly. When designing a system, it may be built from any of the above mentioned solar modules.

Roofing materials are now being manufactured with solar photovoltaics integrated into them. This is a great idea and will bring down the installation costs on new construction. The products are relatively new and it will be interesting to watch this area develop. Generally this technology is based on thin film.

Be careful when using manufacturers data sheets for their solar modules. The module power value generally has percentage tolerances for negative and positive adjustments to this figure. This is a reflection of the imprecision of the solar photovoltaic cell manufacturing processes. Due to this it is common for individual strings of solar modules to either over or under perform in the system. For example, the cloud effect combined with a 10% high performing string of solar modules could produce 165% more than the rated STC current from the string. If your system under performs it may be due to having the bulk of the supplied modules performing at lower than normal power ratings. At least one major manufacturer does not list any negative adjustments on its solar photovoltaic module data sheets, only positive adjustments. This greatly helps with meeting the system performance requirements. It would be good to see other manufacturers follow their lead.

There are some manufacturers that offer 25 year warranties but do not have any product that old. You will need to assess if this is an issue in your decision to use their products. A good question would be to ask yourself if you think that the manufacturer will be in existence in 25 years time to honor the warranty.

Electrically, a solar module is very simple device. It turns light into current at a rate that is proportional to the irradiance it receives. It does this at a voltage that is dependent on the temperature of the solar cells. The solar module will keep indefinitely producing current proportional to the levels of irradiance it receives, even if these levels are much higher than STC due to reflections and lensing. The solar module does this conversion in real time, there is no delay in the conversion process.

A simple approximation of a solar module is that it is a current limited source and that the upper limit for current is entirely dependent on the amount of light received by it!

Due to this, it pays to use the maximum fuse size as recommended by the manufacturer of the solar module.

If the irradience levels are very low at either dusk or dawn then the voltage will be somewhere between 0V and Voc when adjusted for solar cell temperature. At night time it will be at or very close to 0V. The inverter system that the solar photovoltaic modules are connected to knows when to turn on and turn off by monitoring the voltage rise at dawn and the voltage fall at dusk. The inverter system will generally switch on and off at preset voltage levels that have been entered into it during installation.

Solar modules commonly have what is known as "bypass diodes" installed in them. These are installed for shading effects. This enables the other modules in the string to function if one is shaded.

Less common on grid tied systems are "blocking" diodes. One per string can be installed and this prevents reverse currents from flowing into the strings from the system. The string fuses now protect against this by blowing if the reverse current gets too high in a grid tied solar photovoltaic system. It is for this reason that the manufacturer limits the maximum size of the string fuse. If the fuse is above the size recommended and reverse currents occur, then the solar modules in the string may be damaged beyond repair.

If you take a look at the solar module data sheets from the various manufacturers you will see graphs showing the above concepts for their products. Take time to look at the data sheets and understand them, they have a lot of information in them. Don't build a system with their product until you have read and fully understood the data sheet for it.

We will look into the physical sizes of three systems for each solar module type. These will be:

- 5 kWp DC typical residential system

- 250 kWp DC typical commercial system

- 10 MWp DC typical utility system

By comparing the system sizes and components you will be able to see the advantages of each of the different mainstream technologies.

We will not look at the newer solar photovoltaic technologies due to them being in their infancy. Their longevity is currently unknown and most large financing companies will not finance these newer technologies in large installations.

The picture on the following page demonstrates the string concept of solar modules. Solar modules are first wired in series, this is called "stringing", to increase their DC voltage output. Then many strings are connected in parallel to increase the current output of the solar system. Raising the voltage and current is needed for the inverter system to work properly and to increase the efficiency of the system.

The following picture shows the internal diagram of a silicon solar photovoltaic module. As you can see, it is a collection of very large photo diodes wired together into strings. The bypass diodes electrically bypass the internal strings of solar cells if they are either faulty or shaded.

Let us now take a look at each of these technologies and see the differences between typical systems of the same size that are built using the different technologies.

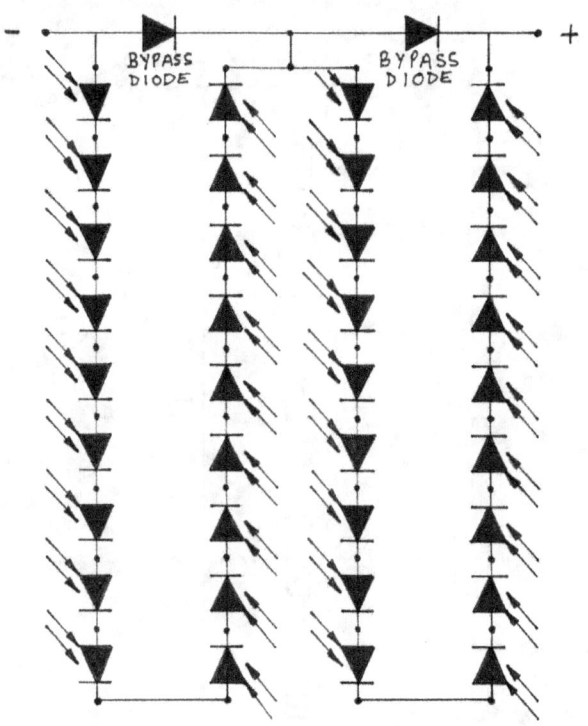

INTERNAL DIAGRAM OF A SOLAR MODULE

A SILICON SOLAR MODULE CONSISTS OF LARGE PHOTO DIODE CELLS AND BYPASS DIODES

11. Mono-crystalline Modules

Mono-crystalline modules are the highest efficiency solar modules available in the general market. What does this mean? Basically an installation constructed with these modules will use less support equipment as these systems use the least amount of solar modules for a given energy requirement. This advantage comes at a cost, as these modules are also the most expensive.

They are made out of a high grade silicon wafers and are very reliable. The mono-crystalline solar cells look black to the eye. Mono-crystalline solar modules have been around for many decades and are very well understood and commonly come with 25 year warranties. Most major manufacturers can show you modules that exceed this age and that are still functioning well today.

Mono-crystalline from most manufacturers is in the 14 to 18 percent efficiency rating. A few have actually manage to exceed 18 percent with over 20 percent being achieved commercially.

Typical Mono-crystalline Solar Module Specifications

- 600V maximum system voltage

- 235W power rating +10% / -5%

- 37V open circuit (Voc)

- 8.6A short circuit (Isc)

- 30V maximum power point (Vmpp)

- 7.84A maximum power point (Impp)
- -0.351% voltage (Voc) temperature coefficient
- +0.053% current (Isc) temperature coefficient
- 15A maximum fuse
- Length 1.64m / 64.6"
- Width 0.994m / 39.1"
- 1.63m² / 17.53 sq ft module surface area

System Calculations

Looking at these dimensional values we get a square meter power of:

$$W/m^2 = \text{rated power} / \text{surface area}$$

$$= 235W / 1.63m^2$$

$$= 144.17 \ W/m^2$$

We dived this value by ten to get the module efficiency per square meter:

$$\text{Module Efficiency} = 14.417\% \text{ per } m^2$$

Assuming a lowest ambient temperature of -10 °C for this area, we need to calculate the maximum voltage for this solar module:

Vmax = (Difference from STC temperature x Voc temperature coefficient x Voc) + Voc

= ((-35) x (-0.351%) x 37V) + 37V

= 41.55V maximum adjusted solar module voltage

Maximum number of solar modules in the strings:

= Rated system voltage / maximum adjusted solar module voltage

= 600V / 41.55

= 14 solar modules maximum in the strings

Wattage of maximum individual string:

= # modules x module wattage

= 14 x 235W

= 3,290W per string

<u>5kWp DC System</u>

Number of modules required in the 5kWp system:

= System wattage / module wattage

= 5,000W / 235W

= 22 Modules

strings = # system modules / # modules in string

= 22 / 14

= 2 strings, each of 11 modules

5kWp system surface area:

= # modules x module surface area

= 22 x 17.53 sq ft

= 385.66 sq ft module surface area

250 kWp DC system

Number of modules required in the 250 kWp system:

= System wattage / module wattage

= 250,000W / 235W

= 1,064 Modules

strings = # system modules / # modules in string

= 1,064 / 14

= 76 strings

Number of 12 input combiner boxes required:

= # strings / # combiner box inputs

= 76 / 12

= 7 combiner boxes required

250kWp system surface area:

= # modules x module surface area

= 1,064 x 17.53 sq ft

= 18,651.92 sq ft module surface area

10 MWp DC system

Number of modules required in the 10MWp system:

= system wattage / module wattage

= 10,000,000W / 235W

= 42,554 rounded up to 42,560 Modules

strings = # system modules / # modules in string

= 42,560 / 14

= 3,040 strings

Number of 12 input combiner boxes required:

= # strings / # combiner box inputs

= 3,040 / 12

= 254 combiner boxes required

10MWp system surface area:

= # modules x module surface area

= 42,560 x 17.53 sq ft

= 746,076.8 sq ft module surface area

Let's make an assumption that when constructed and allowing for spacing between the modules for shading effects that this system will cover 100 acres. We will compare the sizes of our other technologies to this to develop the size of land needed for each technology type.

12. Poly-crystalline Modules

Poly-crystalline solar modules are made from lower grade and therefore cheaper silicon. As such, their efficiency is less than mono-crystalline, but not by that much. A poly-crystalline solar module has a typical conversion efficiency of 13 percent to 16 percent. Their purchase price reflects this lower efficiency and the dollar per watt figure is typically less than mono-crystalline. Poly-crystalline solar cells typically look a metallic blue color to the eye. They are visually a very attractive module to the eye.

Poly-crystalline solar modules have been around for many decades and are a very stable and proven technology.

Typical Poly-crystalline Solar Module Specifications

- 600V maximum system voltage
- 215W power rating +5W / 0W
- 33.2V open circuit (Voc)
- 8.78A short circuit (Isc)
- 26.6V maximum power point (Vmpp)
- 8.09A maximum power point (Impp)
- -0.361% voltage (Voc) temperature coefficient
- +0.06% Current (Isc) temperature coefficient
- 15A maximum fuse
- Length 1.5m / 59.1"
- Width 0.99m / 39"

- 1.485m^2 / 16 sq ft module surface area

System Calculations

Looking at these dimensional values we get a square meter power of:

W/m^2 = rated power / surface area

= 215W / 1.485m^2

= 144.78 W/m^2

We dived this value by ten to get the module efficiency per square meter:

Module Efficiency = 14.478% per m^2

Assuming a lowest ambient temperature of -10 °C for this solar site, we need to calculate the maximum voltage for this solar module:

Vmax = (Difference from STC temperature x Voc temperature coefficient x Voc) + Voc

= ((-35) x (-0.361%) x 33.2V) + 33.2V

= 37.395V maximum adjusted solar module voltage

Maximum number of solar modules in the strings:

= Rated system voltage / maximum adjusted solar module voltage

= 600V / 37.395

= 16 solar modules maximum in the strings

Wattage of maximum individual string:

= # modules x module wattage

= 16 x 215W

= 3,440W per string

5kWp DC System

Number of modules required in the 5kWp system:

= System wattage / module wattage

= 5,000W / 215W

= 24 Modules

strings = # system modules / # modules in string

= 24 / 16

= 2 strings, each of 12 modules

5kWp system surface area:

= # modules x module surface area

= 24 x 16 sq ft

= 384 sq ft module surface area

<u>250 kWp DC system</u>

Number of modules required in the 250 kWp system:

= System wattage / module wattage

= 250,000W / 215W

= 1,168 Modules

strings = # system modules / # modules in string

= 1,168 / 16

= 73 strings

Number of 12 input combiner boxes required:

= # strings / # combiner box inputs

= 73 / 12

= 7 combiner boxes required

250kWp system surface area:

= # modules x module surface area

= 1,168 x 16 sq ft

= 18,688 sq ft module surface area

10 MWp DC system

Number of modules required in the 10MWp system:

= system wattage / module wattage

= 10,000,000W / 215W

= 46,512 modules

strings = # system modules / # modules in string

= 46,512 / 16

= 2,907 strings

Number of 12 input combiner boxes required:

= # strings / # combiner box inputs

= 2,907 / 12

= 243 combiner boxes required

10MWp system surface area:

= # modules x module surface area

= 42,560 x 17.53 sq ft

= 744,192 sq ft module surface area

Comparing this to our mono-crystalline system we get:

Land size = 100 acres x poly surface area / mono surface area

= 100 x 744,192 / 746,078.8

= 99.74 acres

As we can see, there is no significant difference between the area for the mono and poly crystalline modules in this application.

13. Amorphous Silicon (a-Si)/ Microcrystalline Silicon Thin Film

Commonly referred to as the calculator solar cell, due to its use in small solar powered consumer devices. This material has been around for many decades and its properties are well known and understood.

Typical Amorphous Silicon / Microcrystalline Silicon Solar Module Specifications

- 600V maximum system voltage

- 135W power rating

- 249V open circuit (Voc)

- 0.87A short circuit (Isc)

- 188V maximum power point (Vmpp)

- 0.72A maximum power point (Impp)

- -0.3% voltage (Voc) temperature coefficient

- +0.07% Current (Isc) temperature coefficient

- 2A Fuse

- Length 1.409m / 55.5"

- Width 1.009m / 39.7"

- 1.422m^2 / 15.29 sq ft module surface area

System Calculations

Looking at these dimensional values we get a square meter power of:

W/m^2 = rated power / surface area

$= 135W / 1.422m^2$

$= 94.94 \ W/m^2$

We divide this value by ten to get the module efficiency per square meter:

Module efficiency = 9.494% per m^2

Assuming a lowest ambient temperature of -10 °C for this solar site, we need to calculate the maximum voltage for this solar module:

Vmax = (Difference from STC temperature x Voc temperature coefficient x Voc) + Voc

$= (\ (-35) \ x \ (-0.3\%) \ x \ 249V) + 249V$

= 275.2V maximum adjusted solar module voltage

Maximum number of solar modules in the strings:

= Rated system voltage / maximum adjusted solar module voltage

= 600V / 275.2V

= 2 solar modules maximum in the strings

Wattage of maximum individual string:

= # modules x module wattage

= 2 x 135W

= 270W per string

5kWp DC System

Number of modules required in the 5kWp system:

= System wattage / module wattage

= 5,000W / 135W

= 37 rounded to 38 modules

strings = # system modules / # modules in string

= 38 / 2

= 19 strings

5kWp system surface area:

= # modules x module surface area

= 38 x 15.29 sq ft

= 581.02 sq ft module surface area

250 kWp DC system

Number of modules required in the 250 kWp system:

= System wattage / module wattage

= 250,000W / 135W

= 1,852 Modules

strings = # system modules / # modules in string

= 1,852 / 2

= 926 strings

Number of 12 input combiner boxes required:

= # strings / # combiner box inputs

= 926 / 12

= 78 combiner boxes required

250kWp system surface area:

= # modules x module surface area

= 1852 x 15.29 sq ft

= 28,317.08 sq ft module surface area

<u>10 MWp DC system</u>

Number of modules required in the 10MWp system:

= system wattage / module wattage

= 10,000,000W / 135W

= 74,075 rounded up to 74,076 modules

strings = # system modules / # modules in string

= 74,076 / 2

= 37,038 strings

Number of 12 input combiner boxes required:

= # strings / # combiner box inputs

= 37,038 / 12

= 3,087 combiner boxes required

10MWp system surface area:

= # modules x module surface area

= 74,075 x 15.29 sq ft

= 1,132,606.75 sq ft module surface area

Comparing this to our mono-crystalline system we get:

Land size = 100 acres x silicon film surface area / mono surface area

$$= 100 \times 1{,}132{,}606.75 \ / \ 746{,}078.8$$

$$= 151.818 \text{ acres}$$

As we can see, there is a significant difference between the land requirements of the silicon crystalline modules and the silicon film in this application. At this point you would be wise to calculate the initial land costs, installation costs, and the expected operation and maintenance costs for the lifetime of the system.

14. Cadmium Telluride (CdTe) Thin Film

This is the market leader in thin film. It is a coating process that is applied directly to the glass. A nice feature of it is that it does not need bypass diodes, so the solar module junction box is much simpler. This is a relatively young technology and is in widespread use and well understood. It has better high temperature performance than the other technologies.

<u>Typical Cadmium Telluride Solar Module Specifications</u>

- 600V maximum system voltage
- 70W power rating
- 88V open circuit (Voc)
- 1.23A short circuit (Isc)
- 65.5V maximum power point (Vmpp)
- 1.07A maximum power point (Impp)
- -0.25% voltage (Voc) high temperature coefficient
- -0.2% voltage (Voc) low temperature coefficient
- +0.04% Current (Isc) temperature coefficient
- 10A Fuse
- Length 1.2m / 47.244"
- Width 0.6m / 23.622"
- 0.72m^2 / 7.745 sq ft module surface area

System Calculations

Looking at these dimensional values we get a square meter power of:

W/m^2 = rated power / surface area

= 70W / $0.72m^2$

= 97.22 W/m^2

We divide this value by ten to get the module efficiency per square meter:

Module Efficiency = 9.722% per m^2

Assuming a lowest ambient temperature of -10 °C for this solar site, we need to calculate the maximum voltage for this solar module:

Vmax = (Difference from STC temperature x Voc temperature coefficient x Voc) + Voc

= ((-35) x (-0.2%) x 88V) + 88V

= 94.16V maximum adjusted solar module voltage

Maximum number of solar modules in the strings:

= Rated system voltage / maximum adjusted solar module voltage

= 600V / 94.16V

= 6 solar modules maximum in the strings

Wattage of maximum individual string:

= # modules x module wattage

= 6 x 70W

= 420W per string

<u>5kWp DC System</u>

Number of modules required in the 5kWp system:

= System wattage / module wattage

= 5,000W / 70W

= 72 modules

strings = # system modules / # modules in string

= 72 / 6

= 12 strings

5kWp system surface area:

= # modules x module surface area

= 72 x 7.745 sq ft

= 557.64 sq ft module surface area

<u>250 kWp DC system</u>

Number of modules required in the 250 kWp system:

= System wattage / module wattage

= 250,000W / 70W

= 3,572 Modules

strings = # system modules / # modules in string

= 3,572 / 6

= 596 strings

Number of 12 input combiner boxes required:

= # strings / # combiner box inputs

= 596 / 12

= 50 combiner boxes required

250kWp system surface area:

= # modules x module surface area

= 3572 x 7.745 sq ft

= 27,665.14 sq ft module surface area

<u>10 MWp DC system</u>

Number of modules required in the 10MWp system:

= system wattage / module wattage

= 10,000,000W / 70W

= 142,858 modules

strings = # system modules / # modules in string

= 142,858 / 6

= 23,810 strings

Number of 12 input combiner boxes required:

= # strings / # combiner box inputs

= 23,810 / 12

= 1,985 combiner boxes required

10MWp system surface area:

= # modules x module surface area

= 74,075 x 15.29 sq ft

= 1,106,427.47 sq ft module surface area

Comparing this to our mono-crystalline system we get:

Land size = 100 acres x CdTe film surface area / mono surface area

= 100 x 1,106,427.47 / 746,076.8 acres

= 148.3 acres

As we can see, there is significant difference in land use between the silicon crystalline modules and the CdTe film in this application. At this point you would be wise to calculate the initial land costs, installation costs, and the expected operation and maintenance costs for the lifetime of the system.

15. CIGS Thin Film

CIGS comprises of copper, indium, gallium, and selenium in its manufacturing process, hence its name. CIGS thin film looks a streaky greenish black to the eye due to the bath process that is used during the manufacture of the product. CIGS was developed in the nineties and has been a relatively new entrant to the solar photovoltaic field. It is a vacuum deposition process.

Typical CIGS Thin Film Solar Panel Specifications

- 600V maximum system voltage

- 150W power rating

- 96V open circuit (Voc)

- 2.5A short circuit (Isc)

- 70.5V maximum power point (Vmpp)

- 2.15A maximum power point (Impp)

- -0.38% voltage (Voc) temperature coefficient

- -0.06% Current (Isc) temperature coefficient

- 12A Fuse

- Length 1.82m / 71.65"

- Width 1.08m / 42.52"

- 1.97m^2 / 20.65 sq ft module surface area

<u>System Calculations</u>

Looking at these dimensional values we get a square meter power of:

W/m^2 = rated power / surface area

= 150W / 1.97m^2

= 76.14 W/m^2

We divide this value by ten to get the module efficiency per square meter:

Module efficiency = 7.61% per m^2

Assuming a lowest ambient temperature of -10 °C for this solar site, we need to calculate the maximum voltage for this solar module:

Vmax = (Difference from STC temperature x Voc temperature coefficient x Voc) + Voc

= ((-35) x (-0.38%) x 96V) + 96V

= 108.77V maximum adjusted solar module voltage

Maximum number of solar modules in the strings:

= Rated system voltage / maximum adjusted solar module voltage

= 600V / 108.77V

= 5 solar modules maximum in the strings

Wattage of maximum individual string:

= # modules x module wattage

= 5 x 150W

= 750W per string

5kWp DC System

Number of modules required in the 5kWp system:

= System wattage / module wattage

= 5,000W / 150W

= 34 rounded up to 35 modules

strings = # system modules / # modules in string

= 35 / 5

= 7 strings

5kWp system surface area:

= # modules x module surface area

= 35 x 20.65 sq ft

= 722.75 sq ft module surface area

250 kWp DC system

Number of modules required in the 250 kWp system:

= System wattage / module wattage

= 250,000W / 150W

= 1,667 rounded up to 1,670 Modules

strings = # system modules / # modules in string

= 1,667 / 5

= 334 strings

Number of 12 input combiner boxes required:

= # strings / # combiner box inputs

= 334 / 12

= 28 combiner boxes required

250kWp system surface area:

= # modules x module surface area

= 1,667 x 20.65 sq ft

= 34,423.55 sq ft module surface area

10 MWp DC system

Number of modules required in the 10MWp system:

= system wattage / module wattage

= 10,000,000W / 150W

= 66,667 rounded up to 66,670 modules

strings = # system modules / # modules in string

= 66,670 / 5

= 13,334 strings

Number of 12 input combiner boxes required:

= # strings / # combiner box inputs

= 13,334 / 12

= 1,112 combiner boxes required

10MWp system surface area:

= # modules x module surface area

= 66,667 x 20.65 sq ft

= 1,376,673.55 sq ft module surface area

Comparing this to our mono-crystalline system we get:

Land size = 100 acres x CIGS film surface area / mono surface area

= 100 x 1,376,673.55 / 746,076.8

= 184.52 acres

As we can see, there is significant difference between the land area for silicon crystalline modules and the CIGS film in this application. At this point you would be wise to calculate the initial land costs, installation costs and the expected operation and maintenance costs for the lifetime of the system.

16. Solar Module Summary

For residential installations, the supporting infrastructure costs vary little between the different technology types for installation, operation and maintenance.

This starts dynamic starts to change as the system gets larger with commercial and utility installations. As we can see, just an decrease in conversion efficiency of 1% can add acres of land use and supporting infrastructure to a 10 MWp project. The industry is constantly trying to find more efficiency out of their products due to this.

If land use is a significant project expense or is limited, then you probably should be using the more efficient technologies.

Some areas have very cheap land and if this is the case, the more important figures to look at are the supporting infrastructure costs for a larger installation and the ongoing annual operation and maintenance costs for a larger power generation facility. In the dry deserts of the South West USA, these operation and maintenance costs are generally quite low.

Thin film has better performance in low light conditions when compared to the silicon wafer technologies and this may be a factor to consider if the installation area is frequently cloudy.

If the failure rate is the same across all technologies, then the you will be replacing more equipment during the year on a less efficient installation during routine maintenance of the system. Manufacturers appear to be reluctant to release solar module failure rates for their solar module products and on any project, I have become accustomed to assuming a 1% annual failure rate in absence of this information. Unusual and unexpected severe weather conditions may increase this failure rate. Severe weather insurance is recommended to cover these events.

Operation and maintenance costs will be costly for some projects due to:

- Poor site selection

- Poor equipment selection

- Poor wildlife assessment

- Poor build quality

Performing due diligence at the start of any project will keep these problems under control. Ensuring that a highly skilled solar photovoltaic consultant is overseeing large projects from conception through to design and onto construction and commissioning will help to prevent these problems from occurring.

Site selection is probably the most important part of any large solar photovoltaic system project and can be the determining factor in a projects overall success. On any large project it is recommended to install a small version of the proposed system for the first year prior to constructing the main project so that an assessment can be made of how the system will perform in that location and design changes can be incorporated into the main project once all the risks are known. In the large scale solar photovoltaics field it is the tortoise that is generally more successful, try not to be the hare.

Be wary of locations close to the ocean and take the necessary precautions with equipment selection so that high winds, salt spray and corrosion don't become a problem. Salt will deposit itself onto the surface of the modules and if it is excessive, you will be frequently washing the solar photovoltaic modules. Expect the operation and maintenance costs to be higher when installing next to the ocean.

Frame-less solar photovoltaic modules are becoming available and these have better self cleaning properties during rains when compared to framed modules. Frame-less modules are not as strong as framed modules and you will need to assess if this is a problem for your installation or not. Some frame-less modules do not require a ground connection and this can make a significant saving on a large project.

The table on the next page shows the different technologies.

Module Type	Module Wattage	Efficiency	Modules in Strings	# Modules	10 MW Area	Combiner Boxes
Mono	235W	14.42%	14	42,560	100 acres	254
Poly	215W	14.48%	16	46,512	99.74 acres	243
a-Si	135W	9.49%	2	74,076	151.8 acres	3,087
CdTe	70W	9.722%	6	142,858	148.3 acres	1,985
CIGS	150W	7.61%	5	66,667	184.5 acres	1,112

17. Inverters

Inverters convert the DC solar photovoltaic module power into AC power that can be put into the grid. Grid tied inverters come in many different types and sizes. This book is dedicated to grid tie inverters and this is what we will consider. All residential and commercial solar photovoltaic grid tie inverters should comply with UL1741 in the USA. This ensures reliability throughout the various climate zones that the USA has to offer.

There is a lot of discussion as to where inverter technology is going. There are cost savings with using the largest inverter available for large utility installations, but this comes at the expense of poorer MPPT tracking and long DC cable runs which cause power losses. The DC currents are very high in these systems at several thousand amps. Large solar photovoltaic inverter technology (>250 kW) is relatively new in the industry and may not be as reliable as the established small (<10kW) and medium (10kW to 250kW) sized inverters.

AC solar modules are starting to appear and this is at the smallest end of the inverter spectrum. Each solar module has it's own inverter attached to it, commonly called a "micro" inverter. This gives great MPPT performance. This is the ideal configuration for systems that may suffer from:

- Shading
- Have solar modules facing in different directions from each other

- Heavy dirt accumulation

Long term, as the "micro" inverter technology reduces in price I believe that this may become the standard in the industry, even on utility scale installations. The installation is simplified and removes DC from the circuit which can be problematic due to arcing effects which can cause equipment to burn out when it occurs.

Looking at today's technology I currently recommend conventional inverter systems for residential, commercial and utility scale installations. I do not recommend inverters over 250 kW as the DC currents and power flowing through these systems is very large and may cause problems in the solar DC circuit if it is not designed correctly.

Micro inverters can be useful where there is nowhere to mount the traditional inverter on a domestic installation, or if the owner considers an inverter box ugly. The micro inverters can be mounted to the solar panels and hidden from view.

The utility scale inverter systems are still in flux in the USA and currently there is no UL standard for these systems. This will come and for now, the USA utility scale inverter industry is to be regarded as experimental.

When connecting strings to the inverter, do not mix and match solar panel types and connect them to a single inverter, this is a recipe for disaster. If you desire to install the different technology types on the same site, then each solar module technology must have its own inverter system dedicated to it.

Module mismatch in strings is also an issue for the inverter MPPT. Do not connect different wattage modules together in the same string as the higher wattage modules will try and push their current through the lower wattage ones. This will waste power. Each individual solar string needs to have identical wattage modules in it.

Foreign manufactured inverter systems may arrive with the inverter set for a different voltage or frequency of AC electricity. Always check that the inverter system matches the grid system that you are connecting into.

18. Switchgear

Switchgear is widely used in solar photovoltaic installations. A complete line of switchgear has been developed for this application. Specifically, combiner and recombiner boxes are now listed amongst most manufacturers product lines.

Older switchgear was not constructed with solar power generation in mind and may only be rated for electricity flow in one direction. If so, they will have markings on the electrical terminals that say either "load" or "line".

Solar photovoltaic systems feature an AC bi-directional power flow into the grid that reverses at least twice per day. During the day time the solar power system will generate AC energy into the grid and during the night, the power flow will reverse and it will consume a small amount of power from the grid. On cloudy days this reversal will occur frequently during the day time on net zero systems.

As such, all switchgear in the AC circuit on a grid connected solar power system will need to be rated for bi-directional current flow. Due to their bi-directional operation, these electrical systems have no "load" or "line" markings on the electrical terminals.

19. Distribution

Distribution is a feature of the larger commercial installations and the utility installations. Basically distribution covers the medium voltage application which is normally at 24 kV in the USA.

If you are working at this level, it is extremely important that you are competent with working with this type of equipment. The voltages at this level are very unforgiving and death can easily result from misguided activities with this type of equipment. People die every year by electrocution on these systems, it is an occupational hazard. OSHA standard 1910.269 for electric power generation, transmission and distribution details the safety requirements for this type of equipment for the utilities. Work safe and be safe.

Solar photovoltaic systems feature a bi-directional current flow that reverses twice per day. During the day time the solar power system will generate energy into the grid and during the night, the current will reverse and it will consume a small amount of power from the grid. As such, all distribution equipment in a solar photovoltaic power system will need to be rated for bi-directional current flow.

Distribution generally covers the following items

- Transformers
- Distribution poles

- Medium voltage switchgear
- Relays
- Harmonics

Transformers

Dry type transformers are generally used in Solar photovoltaics and have very high efficiencies of over 99% conversion. Some people are still using oil filled transformers and the operation and maintenance costs are higher with these due to needing more attention and care.

Distribution Poles

Many installations use distribution poles to connect into the point of interconnection (POI) at the grid. It is a cheaper way of distribution when long distances are involved. For short distances of less than about 1 mile there really isn't any difference between using distribution poles or going underground, if the underground route is easy to put ducts and conduits into.

Medium Voltage Switchgear

Medium voltage switchgear is generally of the type:

- Gas insulated
- Vacuum insulated

The choice of which type to use is generally driven by costs. The higher power switchgear is generally vacuum insulated.

Protective Relays

Protective relays control the tripping mechanisms on the switchgear. They are set up according to the desired fault levels that the designer has calculated for each switch on the system.

Harmonics

Inverter systems cause harmonics. Normally, total harmonic distortion (THD) should below 3% and is quite acceptable to feed power into the grid at these low levels.

20. Ancillary Systems

These systems are not essential and are generally installed on larger sites, particularly if they are unmanned:

- Weather Station
- Security System
- Video Surveillance
- Metering
- Inverter Monitoring
- String Monitoring
- Monitoring Services

Weather Station

Many solar sites install weather systems, particularly if they are unmanned. They can provide useful information that can be used to verify the power output of the system. Most solar site weather systems include the following sensors:

- Reference solar cell with temperature
- Irradiance sensor
- Solar module temperature sensor
- Ambient temperature
- Humidity
- Wind speed and direction

- Rain gauge

Lets look at each of these in detail:

Reference Solar Cell With Temperature

This reference cell will be made from the same solar cell type as the installed solar modules contain. It is a factory calibrated solar cell. It has a temperature probe attached to it to record the cell temperature. This allows the cell reading to be adjusted to the correct value.

Irradiance Sensor

This is used to record the irradiance levels and is sometimes referred to as a pyranometer.

Solar Module Temperature Sensor

The solar module temperature sensor is attached to the rear surface of one of the installed solar modules. This allows the operating temperature of the solar module to be known and for the solar module output to be adjusted for temperature effects.

Ambient Temperature

The site ambient temperature is recorded for historical purposes.

Humidity

The site humidity is recorded for historical purposes.

Wind Speed and Direction

The wind speed and direction can be used to assess the cooling effects from this on the solar modules.

High wind speeds are a concern for system damage from flying debris. After a high wind speed event has been detected, a site visit should be arranged to check on the equipment.

Rain Gauge

It is useful to know the historical rainfall, so that you can schedule cleaning. If you have a site with consistent rainfall, then it is unlikely that you will need to clean the solar modules regularly. If you have a site that only gets sustained rainfall on one or two occasions per year, you may want to schedule solar module cleaning more frequently.

If any part of the weather system fails, then it is usually replaced. Weather system parts are generally regarded as consumables due to the harsh environment that they work in.

Security System

On remote sites a security system is usually installed. The security system will usually use beam break sensors within the caged areas of the solar site. Keep these beam break sensors clean to ensure that false alarms do not occur.

Thefts have not been a significant problem in the industry yet. It helps that the solar modules are firmly attached to their mounting structure and that tools are needed to remove them. Generally most thieves are after the copper wiring. This is why it is important to always check the grounding systems of your installations, as they may have been stolen.

As public understanding of solar photovoltaic power systems increases and demand for the solar modules goes mainstream, this may develop. It is a good idea to install the security system during construction of the project to keep your insurance premiums down on the system.

Video Surveillance

On remote sites usually internet based video surveillance cameras are installed with the pan, tilt and zoom (PTZ) feature. These will be low light level cameras that can work under moonlight. They are mounted in weather resistant housings.

Metering

The electrical meter for the solar photovoltaic system can be connected to the internet. This is useful for billing purposes if the energy is being sold to a customer. It also is very useful for detecting a problem at a remote site. If the site is not producing energy, then it can be quickly detected by monitoring the meter.

Inverter Monitoring

Most inverters have the capability to be monitored and controlled. This is useful for power factor control and for detecting inverter faults. If the inverter has power factor control, then it can be used to provide power factor control to the grid that it is connected to. The utilities like to have this feature if it is available.

The inverters have many fault codes and it is useful for a remote site to know the fault code in order to schedule a repair visit.

A nice feature that the inverters have is the ability to monitor the DC inputs to them. This is commonly referred to as "zone monitoring". When monitoring at the zone level, it is easier to detect problems with the solar module strings by comparing the zones to each other. If one zone is consistently reading less than another equally sized zone, then you will need to schedule a repair visit to that zone.

String Monitoring

Monitoring can be done at the string level. On a residential system this is normal for the inverter to be monitoring at this level due to the small number of strings connected to it. On commercial and utility systems, this would be installed in the combiner boxes.

On utility systems with several thousand solar photovoltaic strings, this is a lot of monitoring and as such, it is also expensive to install and maintain. It is nice to have, but not really needed. Good inverter zone monitoring should be able to detect an individual string failure.

Monitoring Services

There are many companies now offering monitoring services for solar power generation systems. While not so useful for residential and commercial applications where there will be people present to check on the system, they are extremely useful for remote sites.

The monitoring center will send out automatic email notifications to you when problems are detected on site. This enables a response to be arranged quickly before too much energy generation is lost. On a large solar photovoltaic power generation system, it doesn't take much time of the system being out of service before thousands of dollars of energy generation has been lost.

These monitoring services are invaluable on large generation systems and their expense is far outweighed by the revenue saved by keeping the solar power generation system in full service.

21. Power Purchase Agreements (PPA)

Many solar installations are built using power purchase agreements(PPA) that are used to finance the project. A PPA is quite a simple idea. A loan to finance the project is taken out by the company that builds it. It is built on land that is owned by the customer. Once in operation and producing energy, that energy is sold to the customer who has agreed to purchase all energy produced by the system over twenty years. The sales of energy during the twenty years are used to pay for the operation and maintenance costs of the system and the original loan interest and principal.

After about ten years the loan is repaid in full and all future sales of energy produce profit at that point after routine costs. After the twenty years is up, the customer can either choose to enter into another PPA for energy supplied, have the system removed from their property or purchase the system.

The PPA is a great concept and when a system performs as planned or even better, they are a great way of building solar photovoltaic systems. Due diligence at the planning stage ensures this.

22. Site Licensing Agreements (SLA)

Site licensing agreements (SLA) are the partner to the power purchase agreement (PPA). If you are building on a piece of property that is owned by someone else, you will need one of these. It is basically a contract for land use for the duration of the project.

It is important that the SLA is well written and incorporates the following:

- Term of land use

- Unrestricted access for the duration of the project

- No impacts to the solar photovoltaic power generation system from future surrounding developments (generally shading & dirt problems)

- Energy costs

- Guaranteed power purchasing by the customer

- Customer maintains the point of AC interconnection in service at all times

- Customer pays compensation if the point of interconnection is not available for power production.

- End of lease needs to be well defined for the options to enter into another PPA, purchase the system, or have the system removed.

23. Summary

Solar photovoltaics is a very reliable technology. The field is set for growth and systems should become commonplace with the support of the government incentives. When designed and installed correctly on a good site, solar photovoltaic power systems will reliably produce energy for many years with minimal maintenance.

In the future it is quite possible that every new home constructed will come with a solar photovoltaic power system as standard. Installing solar photovoltaics during the construction phase of a home is the most cost effective way to adopt the technology.

Many companies have solar photovoltaic roofing products on the market and in development to meet the future needs of modern, self contained buildings. Net zero is probably the future of green construction, where a building will produce its own energy needs with an annual energy use of zero from the utility grid.

With the development and adoption of AC solar module technologies, solar photovoltaics will not need any specialized knowledge to install it and any electrician will be able to work with it. AC solar photovoltaic modules will be just like an appliance that will plug into the house electrical system. Indeed, in the future you may well purchase your solar photovoltaic modules at the appliance shop.

The smart grid is in development to allow these distributed generation systems to be effectively managed by the utilities. The electrical grid will be quite different in the future with power generation taking place on the majority of consumer premises with the utility grid supporting it when sufficient power cannot be generated due to climatic conditions.

The utilities will continue to build large solar photovoltaic power plants and they will keep getting larger with each generation of plant. The utilities are currently learning how to construct solar photovoltaic systems that work well with their existing infrastructure. Today's utility systems are paving the way towards the very reliable solar photovoltaic power generation systems of the future.

It is an exciting time in the solar photovoltaic industry and I hope that this book has given you an overview of where the technology is today and where it may be heading in the future. By reading this book you will now be prepared for entering the field of solar photoltaics and will know how to put systems together.

I recommend that further reading is done in addition to this book. The reference section is a good source for further information, as are my other books on the subject. This book is not a substitute for a formal training course on solar photovoltaics and this would be a great next step in the learning process.

I hope that you have enjoyed this book and welcome to the wonderful world of solar photovoltaics!

24. References

- NFPA National Electrical Code 2005

- NFPA National Electrical Code 2008

- IEEE/ANSI-C2-2007 National Electric Safety Code

- Occupational Safety and Health Administration www.osha.gov

- National Renewable Energy Laboratories www.nrel.gov

- United States Department of Energy www.energy.gov

- Solar Photovoltaic Design for Residential, Commercial and Utility Systems by Steven Magee

- Solar Photovoltaic Operation and Maintenance for Residential, Commercial and Utility Systems by Steven Magee

- Solar Photovoltaics for Consumers, Utilities and Investors

25. Author Contact

Steven Magee,

3618 S. Desert Lantern Road,

Tucson,

AZ 85735

USA

I hope that you found the book informative and please let me know about any questions or comments about the book.

I am a consultant on new solar photovoltaic projects, solar photovoltaic system troubleshooting, solar photovoltaic training, and solar photovoltaic investing for financial companies. Please feel free to contact me for any help or assistance in these areas.

You may find my other books useful:

- Solar Photovoltaics for Consumers, Utilities and Investors

- Solar Photovoltaic Design for Residential, Commercial and Utility Systems

- Solar Photovoltaic Operation and Maintenance for Residential, Commercial and Utility Systems